Amazon Fire Phone User Manual: Guide to Help Unleash Your Smartphone Device!

By Shelby Johnson

www.techmediasource.com

Disclaimer:

This eBook is an unofficial guide for using the Amazon Fire Phone product and is not meant to replace any official documentation that came with the device or accessories. The information in this guide is meant as recommendations and suggestions, but the author bears no responsibility for any issues arising from improper use of the Amazon Fire Phone. The owner of the device is responsible for taking all necessary precautions and measures with their device.

Amazon Fire Phone is a trademark of Amazon and/or its affiliates. All other trademarks are the property of their respective owners. The author and publishers of this book are not associated with any product or vendor mentioned in this book. Any Amazon Fire Phone screenshots are meant for educational purposes only.

Content

___

—

Introduction

The Fire Phone is the first smartphone from Amazon. The company reportedly spent five years creating the phone making sure it could provide you with a superb smartphone experience.

The new phone from Amazon comes at a time when many people have a smartphone, but Amazon managed to include features in its new device that are completely unique to the class, which makes it a great new phone to own especially if you are an Amazon Prime customer or plan to become one. Of course initially, every person who buys a Fire Phone will get an entire year of Amazon Prime service included in the phone's price.

While some people may think that the Fire Phone's features like Firefly make it big cash register for Amazon purchases, others who already purchase often from Amazon will love the added convenience the phone offers.

I own and use this new phone, and I created this guide to help you get the most out of your new Fire Phone. Read on to unleash the power of your Amazon Fire Phone!

Fire Phone Specs

Amazon did not skimp on the Fire Phones specs. This new smartphone features a 4.7-inch dynamic display with dynamic image contrast, which allows for easier outside viewing. It also has a 2.2 GHz quad-core Snapdragon 800 processor and 2GB of RAM to allow for quick app loading and smooth multi-tasking with your phone. It runs on Fire OS 3.5.0, which is Amazon's version of the latest Android operating system. The phone's beautiful Adreno 330 graphics allow for immersive gaming and videos.

The Fire Phone has plenty of battery life too with up to 285 hours of standby time, up to 22 hours of talk time, up to 65 hours of audio playback, and up to 11 hours of video playback. You will not have to worry about battery conservation until you have used your Fire Phone a lot during a day.

Your Fire Phone measures 5.5" x 2.6" x 0.35" and it weighs just 5.64 ounces. You get all this and more with your Fire Phone, and Amazon provides a one-year limited warranty.

What's in the Box

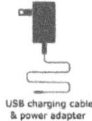

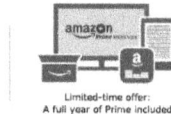

The Fire Phone comes with everything you need to get started right in the box. When it arrives and you open the box you will find the following:

- Fire Phone.
- USB charging cable and power adapter.
- Premium, tangle free headphones with a mic and remote.
- A small quick start guide.
- For a limited time, 1 year of free Amazon Prime service.

What's in the box

| Amazon Fire Phone | USB charging cable & power adapter | Premium headphones with remote & mic | Quick Start Guide | Limited-time offer: A full year of Prime included |

After you set up and start using your Fire Phone, you may find you want some additional accessories, so I have covered my favorite Fire Phone accessories later in this guide.

Fire Phone Layout

It is important to familiarize yourself with the Fire Phone layout before getting started. The following image shows a layout of the device, including the buttons on the side, top and bottom panel of the phone that you'll be using for power, volume, going to the home screen, or for activating either the camera as well as Firefly features of the phone. In addition, you can see where the micro USB cable is connected to the device, as well as where the SIM card is stored in your phone for future reference.

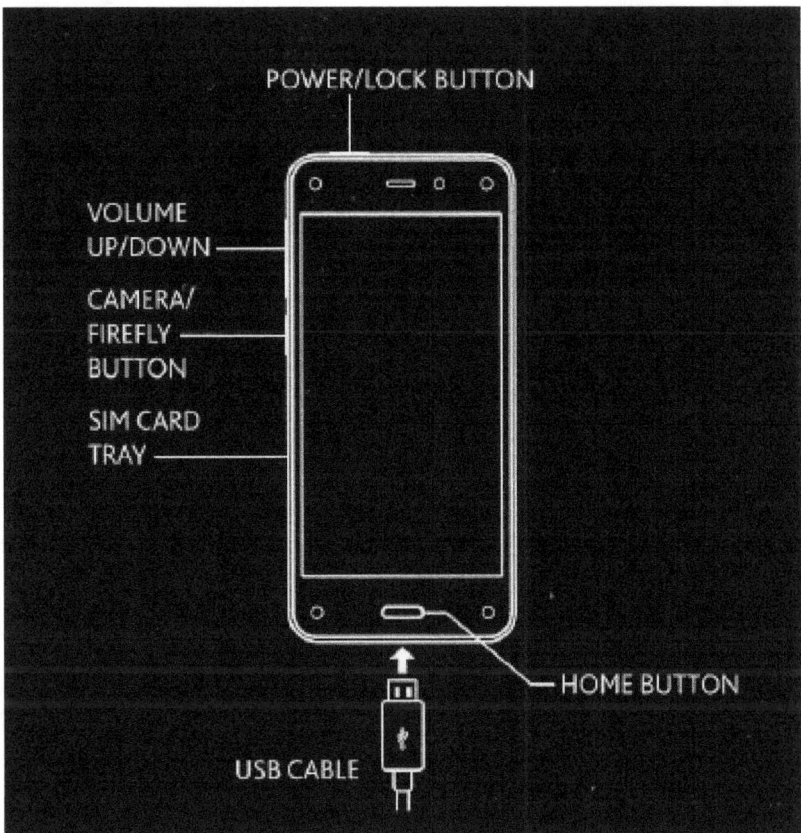

The next sections describe the phone's screens and settings.

Fire Phone Screen Display & Navigation

The Fire Phone is a unique smartphone, so there are several ways to get the screens to display. The sections below describe these screens and the specific movements that open them.

There are four main screens on the Fire Phone. They are the Center Screen, Left and Right Panels, and Quick Actions Panel. There is also the Carousel view and the Apps grid.

Tip: *To return to the previous screen from any screen on the phone, simply swipe up from the bottom.*

Center Screen – Shows your activities and content. This screen can either be viewed in the Carousel view, or you can swipe up from the bottom of the screen, and view it from the Apps grid view.

Left and Right Panels – The Left Panel (pictured in the image below) brings up a menu of Apps, Games, Web, Music, Videos, Photos, Books, Newsstand, Audiobooks, Docs, Shop, and Prime. While the Right Panel transforms based on what you are doing on the phone at any given time. On home the Right Panel gives you up to the minute personalized info. While listening to music right panel might show you X-Ray information about the lyrics. When you send a text message, the Right Panel has photos you can attach to the text message.

```
11:10 PM          📶  Emergency Calls  80% 🔋

APPS    A new paid app for
        free every day

GAMES                              wser

WEB

MUSIC                              ler

VIDEOS  Watch 200,000
        Movies & TV episodes

PHOTOS

BOOKS                              k

NEWSSTAND

AUDIOBOOKS                         ideo

DOCS

SHOP    Millions of items,         itor
        delivered to your door

PRIME   Free two-day shipping
        & more
```

To open these panels, simply tilt the phone's right or left edge towards you to open either the right or left panel, respectively. To close the panel, simply tilt it towards the opposite edge of the panel you have open.

Quick Actions Screen – The Quick Actions Screen has Flashlight, Mayday, Airplane Mode, Search, and Settings. To access it, simply swipe down from the top of the screen, or swivel the phone with your wrist.

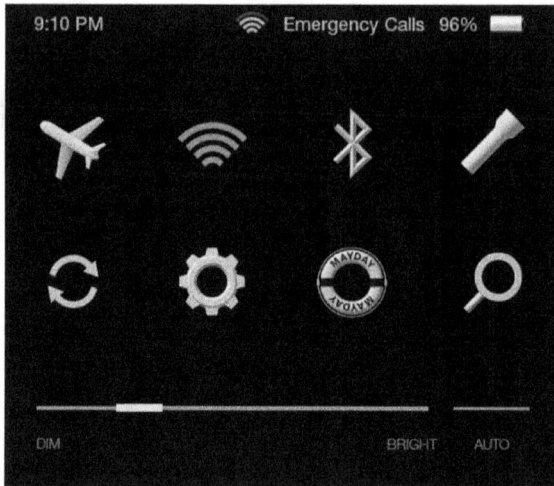

The image above shows the Quick Actions Panel. This is an easy way to access Settings for the Fire Phone.

Peek – You can get a "peek" when you slightly tilt your Fire Phone to reveal further info on the app or screen you're on. For example, in the Maps app, tilt your phone just a bit to either side and it will bring up relevant locations on the map such as nearby restaurants with reviews and ratings.

Settings

The Settings screen is where you will modify many different aspects of the Fire Phone. You will most likely often access Settings for many different reasons.

Settings

Expand All

Search Settings 🔍

Wi-Fi & Networks

Display

Sounds & Notifications

Applications & Parental Controls

Battery & Storage

Location Services

Lock Screen

Keyboard

Phone

My Accounts

Device

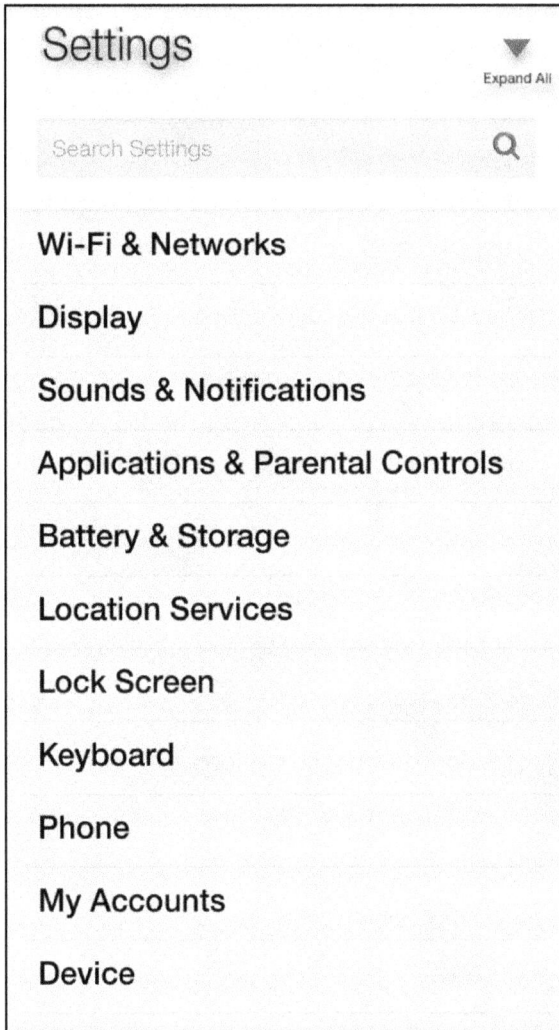

The Fire Phone Settings categories are as follows:

- Wi-Fi & Networks
- Display
- Sounds & Notifications
- Applications & Parental Controls
- Battery & Storage
- Location Services
- Lock Screen
- Keyboard

- Phone
- My Accounts
- Device
- Voice
- Help & Feedback

To access Settings on the Fire Phone, do the following:

- You can access settings on your phone by swiveling the phone or swiping down from the top of the screen to open the Quick Actions Panel. Tap the Settings icon.
- You can also access settings from the Home screen. Simply tap the Settings icon in the carousel or app grid views.

Getting Started/Initial Setup

The following sections will help you get started with the Fire Phone.

Activating Wireless & Data Service

If you don't already have service activated for your Fire Phone, make sure you have activated your phone with a new wireless service plan. It is best to refer to any instructions or guides you received with your Fire Phone to do this through the appropriate phone and data service provider.

As of the original publication of this guide, AT&T is the exclusive service provider for the Fire Phone. Make sure you have your SIM card number and IEM number from the Fire Phone box label before you call or go online to activate wireless service.

Note: *For individuals who purchased the phone without service on Amazon's website, your phone includes a pre-installed SIM card. You can choose to activate service online. When you activate, simply choose one of the available plans including monthly Pay as you Go plans. Be sure you have the SIM card number and IMEI number from your Fire Phone box label available when activating service. Refer to all included instructions with the device for activating service.*

Setting up the Fire Phone

Once you have your device out of the box plug the micro usb cable into your Fire Phone and the wall charger, then plug the wall charger into an available socket to fully charge your smartphone.

When the phone is charged, power your phone on by holding down the power button on top of the Fire Phone. The phone will show a screen with an Amazon logo and then a Fire logo screen for a bit of time as it starts up. Soon you can begin to set up the phone.

1. First, Select your language (English or Espanol).
2. Select your Time Zone from those listed.
3. Connect to your Wi-Fi network (make sure to have any login passwords needed to connect to your preferred Wi-Fi network).
4. Verify or connect to your Amazon account (you will need your Amazon account login info to do this).
5. Confirm your Amazon account and agree to the terms by clicking "Next."
6. Enable Location Services – This will allow Maps, other apps and Amazon to use your location and related info to provide more relevant info for you on the phone. (Choose "No Thanks" or "Enable").
7. Enable Backup – This will backup Fire settings, notes, search history, bookmarks, communications, call history and more in the Cloud in case you need to restore a Fire phone later. ("No Thanks" or "Enable").
8. Connect your Social Networks. Login to your Facebook and/or Twitter accounts here to integrate these accounts with your Fire Phone. Supply your login email or username and passwords for either or both social networks to connect them.

After completing the initial setup, there is a Fire welcome "quick interactive guide" which will help you better understand how to use your new phone. The interactive video tutorial will show you the basics of using your smartphone, including motion gestures and how to find your way around the phone menus. You can also opt to skip this step by clicking on "Skip" on the bottom left corner area.

How to De-register Fire Phone from Amazon Account

You may have given the Fire Phone to someone else as a gift, or you may simply want to de-register it from one account and add it to another. In order to de-register the Fire Phone, complete the following steps.

1. Go to Amazon.com.
2. Hover your mouse over "My Account."
3. Click on "Manage Your Content and Devices."
4. Click on the "Your Devices" tab.
5. Click on the Fire Phone listed under "Your Devices."
6. Click on "Device Actions" and choose "Deregister."

After you de-register a device from one account, it is ready to be registered to another account.

How to Transfer Contacts from Old Phone

If you are changing to the Fire Phone from another type of smartphone like an iPhone and Android, you can transfer your contacts from your old phone.

Both types of phone have different instructions for how to transfer. The following sections will explain how to transfer from the different types of phones.

How to Transfer iPhone Contacts

You can transfer your iPhone contacts to the Fire Phone. The following steps describe the process.

1. From your iPhone, tap Settings > iCloud, and ensure syncing is on for Contacts.
2. Disable iMessage so you can send text messages to your Fire phone. To do this, tap Settings > Messages, and then turn iMessage off on your iPhone.
3. Note: You must to turn off iMessage on all other iOS or Mac devices tied to your Apple ID.
4. Download the AT&T Mobile Transfer app on your iPhone from the Apple App Store.
5. Download the AT&T Mobile Transfer app to your Fire phone from the Amazon Appstore.
6. Open the app AT&T Mobile Transfer on both phones and follow the instructions to link devices.
7. On your Fire phone, select the content you want to move.
8. Tap Transfer.

Note: *You can move photos and videos through this app, but Amazon suggests using the Amazon Cloud Drive Photos app, which is described in another section of this guide.*

How to Transfer Android Contacts

Transferring your Android Contacts to the Fire Phone takes a few simple steps as long as your phone is on the supported phone list.

If your phone is supported, complete the following steps to transfer your contacts from your Android phone.

1. Download AT&T Mobile Transfer on your Android phone from the Google Play store
2. Download AT&T Mobile Transfer on your Fire Phone from the Amazon Appstore.
3. Open the app on both phones and follow the instructions to link your devices.
4. On your Fire Phone, select the content you want to transfer from your Android.
5. Tap Transfer.

Note: *Only contacts stored directly on your Android will transfer. Those saved online will not transfer via the above method.*

Note: *You can move photos and videos through this app, but Amazon suggests using the Amazon Cloud Drive Photos app, which is described in another section of this guide.*

How to Set up E-mail

The Fire Phone's e-mail app supports most POP, IMAP, and Exchange accounts, including Gmail, Outlook, and Yahoo! Mail. These apps will automatically populate server settings during setup for most popular e-mail providers.

However, if you are not using one of these popular e-mail providers, you may have to manually set up your e-mail account. To do this, you will need your provider's e-mail settings. Information for how to set up POP3, IMAP, and Exchange accounts if you receive the Advanced Setup screen is located after the following basic e-mail setup instructions.

1. From the Email app, access the left panel, and then tap Settings.
2. Under Accounts, tap Add Account.

3. Enter your e-mail address, and then tap Next.

NOTE: *If your e-mail account isn't recognized, you will see the Advanced Setup screen.*

For POP3 or IMAP accounts:

1. Tap POP3 or IMAP at the top.
2. Tap POP3 Server or IMAP Server, and then enter the server URL. For POP3, the URL is pop3.abc.com. For IMAP, the URL is smtp.abc.com.
3. Tap Username, and then enter your username.
   Tip: *Your username is typically your e-mail address.*
4. Tap SMTP Server, and then enter the outgoing SMTP server name (e.g. smtp.abc.com).
5. Tap Security Settings and Ports. Verify that your Incoming and Outgoing Security Settings are correct. You should contact your e-mail provider for the correct security settings.

For Exchange accounts:

1. Tap Exchange at the top.
2. Tap Exchange server, and enter the server URL <exchange.mycompany.com>.
3. Tap Username, and enter your username.
   Tip: *Your username is typically your e-mail address without the domain.*
4. Tap Security Settings and Ports. Verify that your Incoming and Outgoing Security Settings are correct. You should contact your e-mail provider for the correct security settings.

Note: *Exchange e-mail accounts often require that you accept device policies as determined by your system administrator. If required, you will be prompted to accept policies such as a device PIN/Password and encryption. Simply follow the prompts to complete your account setup.*

Once your e-mail account is configured, you can tap Go to Inbox to immediately view the account you just added. If you would rather add additional accounts at this time, simply tap Add Another Account to repeat the steps above to add additional e-mail accounts.

Fire Phone Features

The Fire Phone has some excellent features. Amazon really went all out in developing its first phone, so this new phone offers some features that are not available in any other smartphone on the market. Check out more details and helpful tips on some of the phone's major features below.

Home Carousel

At the top of your home screen are a series of large icons referred to as the "Carousel" that you can scroll through from left to right and vice-versa. The Carousel allows you to scroll, scan, and take action without even having to open an app or leave the carousel. Using the Carousel, you can scan and delete emails, check out when your next appointment is, locate your most recent photos, quickly access your most visited websites, and read your most recently purchased eBooks, and much more. The next image the Weather app and Firefly app on the Carousel.

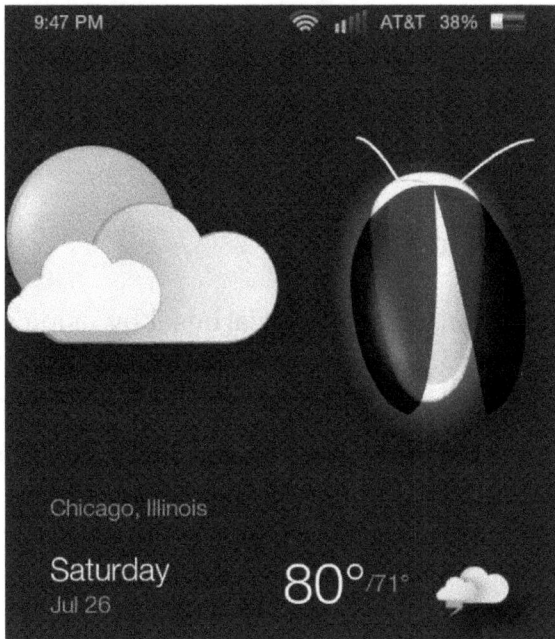

Chicago, Illinois

Saturday
Jul 26

80°/71°

You will notice that various apps, websites and other features that you use will show up on the Carousel. On some of the Carousel items you scroll to, you will see various notifications below them. Examples are upcoming weather reports, your most recent text messages, notes, or other important info from the apps and features you have used.

Your latest available Kindle eBooks will also be available to scroll through on the Carousel along with any music or video content you may have watched or purchased recently.

The Carousel is also customizable, so you can have the things that are most important to you on the Carousel. Using Pin to Front, you can pin e-mail, photos, or specific apps to the front of the carousel, which provides you with quick access.

It is convenient to have this type of access just using the Home Carousel. Many other smartphones do not offer this type of home screen convenience.

Note: *To remove an item from your Carousel hold down on the item's icon image until you see "Remove from Carousel" pop up. Tap that to remove the item from your carousel.*

How to Pin to Front

You can further customize your Carousel by using an option called Pin to Front. In order to pin any app to the front of your Carousel to easily locate it in moments, do the following:

1. Press and hold the icon of the app you want to Pin to Front.
2. You will see two or three options including Pin to Front, Remove from Carousel, and Remove from Device (this option is for third party apps only).
3. Tap Pin to Front to include app on your Carousel.

Note: *To arrange the apps on your Carousel, press and hold the app icon, and tap "move to front."*

Firefly

Firefly is an exciting feature on the Fire Phone that not only recognizes printed text allowing you to make calls, save contacts, or visit websites at the touch of a button, but it also recognizes movies and TV shows just by hearing part of the video's dialog.

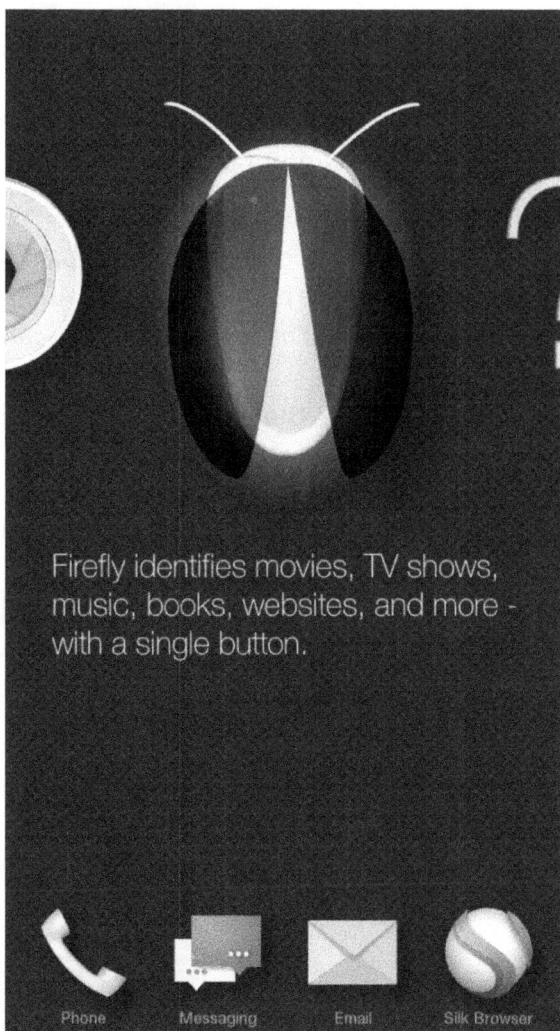

Firefly identifies movies, TV shows, music, books, websites, and more - with a single button.

Phone   Messaging   Email   Silk Browser

The amazing Firefly feature also recognizes songs giving you access to the artist information, albums, similar music, and more. You can enable Firefly using apps like iHeartRadio. Firefly can also help you find tickets to the artists' next show using StubHub.

Additionally, Firefly can recognize millions of household products and will provide product information, add it to your Wish List, or allow you to purchase the product directly from Amazon's website.

How to Use Firefly

Tap on the Firefly icon on your apps or Home Carousel OR press and hold the side camera button on your Fire Phone.

- The Firefly app will open up with camera view. You can hold your Fire Phone to aim its camera at a UPC barcode of a product to search and shop for it online via Amazon. Keep in mind not all products will be identifiable. You can also scan QR codes.
- To identify music, tap on the music note icon at the top of the Firefly app. Hold the Fire Phone within a good range of the speaker playing the music. After several moments, Firefly should identify the song, artist and album the music is from.
- For TV or movies, tap on the TV icon at the top of the Firefly app screen. Hold the Fire Phone up aimed at the screen for several moments. The Firefly app will identify the movie or TV show that is on based on recognizing the sounds and/or picture.

X-Ray

When Amazon introduced X-Ray for eBooks and textbooks, the world rejoiced, thanks to the ability to look up pertinent information about characters, places and events without leaving the current page they are on. The same rejoicing happened when X-Ray became available for music. The Fire Phone has X-Ray for all three available, which is also a feature other smartphones do not offer.

As a song plays, X-Ray will list the artist and title, along with the lyrics. The lyrics will be displayed on the right side of the screen, and will scroll line by line as the song plays. Now you never have to wonder what your favorite artist is saying, or look to another source for the information. It is all right in front of you with X-Ray for Music.

Meanwhile, X-Ray for Video shows you an actor's name, biography, other roles, and scenes without leaving the movie. It delivers the ever-popular IMDb app information within the movie. To access this feature on a movie that offers X-ray, all you need to do is tap the screen during a scene and it will pop-up information such as which actors and actresses are appearing and more! You can also activate Firefly while watching a movie or show on your TV and aim the camera towards the screen to identify the related details with X-Ray as seen in the screenshot.

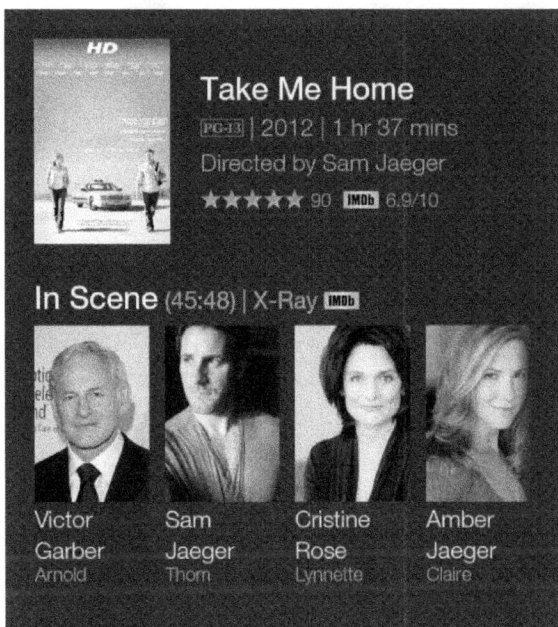

With X-Ray for eBooks you can trace characters, places, and ideas as you read an eBook. The feature provides the data using information from Shelfari, YouTube, and Wikipedia, providing an excellent companion for readers.

How to Use X-Ray

The X-Ray information on the Fire Phone is located in the Left Panel. To use X-Ray for X-Ray enabled eBooks, music, or videos, simply tilt the left edge of your phone towards you, or swipe over from the left to open the Left Panel. X-Ray details will be located in the Left Panel.

Dynamic Perspective

Dynamic Perspective is unique to Amazon's Fire Phone. This revolutionary smartphone feature allows you to tilt, auto-scroll, and swivel using just one hand, which is convenient for a phone. The Dynamic Perspective also allows for one-handed reading.

In addition to convenience in navigating, Dynamic Perspective also provides you with an immersive experience using apps and games with your Fire Phone. In games, you will be able to see from the character's perspective and actually peek around obstacles. Meanwhile, in the maps app, you will be able to see Yelp ratings, and StubHub on the Fire Phone will provide you with views 90 degrees to the right and left of your seat.

Mayday

If you need any help with your Fire Phone, all you have to do is press the "Mayday" button. Amazon has live help available 24 hours a day, 365 days a year. Mayday calls are typically answered in a mere 15 seconds. This feature is the only one of its kind to be offered in a smartphone.

The availability of a Mayday representative is great because you do not have to sit on hold waiting for the right answer, and you will have a physical presence with you to show you through the problem without issue.

You will receive a live customer service rep on streaming video who you can see, but can't see you so there is no more searching for a customer service representative verifying description of an item on (or off) screen, and there are no more technical terms necessary to make you feel like a fool for owning such an awesome device.

This option is available at all times, so use the dedicated Mayday button to activate the feature and allow someone at Amazon to help you right away by literally pointing at the problem – or the solution.

How to Use Mayday

To use Mayday, do the following:

1. Swipe down from the top of the screen to open the Quick Actions panel.
2. Tap the Mayday button.
3. Tap Connect, which is an orange button in the center of the screen.

You will see a Mayday representative in the lower right corner of your screen, and you can begin talking to the representative about your problem or issue.

Second Screen

Much like Apple has Airplay, which allows you to turn your iPad into a remote for your big screen with an Apple streaming device, the Fire Phone has followed suit by turning your smartphone into an entertainment hub that "flings" content onto a larger screen, to view with a group. This content can be pictures, videos, movies, television programming, or even your email inbox.

All you need is a compatible device such as the Amazon Fire TV, a Samsung television, or a PlayStation 3 or 4 for streaming purposes, and you are all set. The best part is, you can continue to work on your phone while the television displays alternate content, so you can check emails or look for content on the web as you desire, while enjoying the programming remotely.

Immersion Reading

There is a great way to enjoy books on the Fire Phone, and it is called Immersion Reading. When enabled, an audiobook will read the words, as the screen highlights them. This feature is perfect for beginning readers, allowing them to follow along within the text of the book, while the voiceover reads aloud.

ASP

ASP, or Advanced Streaming Prediction, is a feature of the Amazon Fire phone that predicts your viewing habits and automatically buffers the programs it believes you will want to view, so there is no delay in watching the television shows, movies, or videos when you click on them. Of course, the more you use Fire TV, the more it will know about your viewing habits, allowing this feature to work with precision.

Whispersync

Whispersync for Voice is cutting edge technology that allows you to toggle between reading a Kindle book and listening to its companion audiobook without losing your place. It will also remember your position, and keep your notes and bookmarks.

When used for games, movies and books, this technology will sync each device you are enjoying the entertainment on, placing you directly where you left off on the other.

Whispersync works with Kindle e-readers and reading apps, Amazon Instant Video, and some GameCircle titles.

Fire Phone Accessibility Features

Your Fire Phone comes with accessibility features for users who have vision, hearing, or mobility impairments.

Note: *Turning on an accessibility feature may cause your phone to function differently (such as different gestures for navigation).*

The Fire Phone comes with the following accessibility options:

- Screen Reader
- Explore by Touch
- Screen Magnifier
- Power Button Ends Calls
- Low Motion Mode
- Spoken Caller ID
- Convert Stereo to Mono
- Hearing Aid Mode
- TTY Mode
- Closed Captioning

How to Turn On and Off Accessibility Features

To turn on accessibility features complete the following:

Swipe down from the top of the screen to open the Quick Actions panel.

1. Tap Settings.
2. Tap Device.
3. Tap Manage Accessibility.
4. Select your accessibility options.

To turn off accessibility features do the following:

Swipe down from the top of the screen with two fingers to open the Quick Actions panel.

1. Tap Settings.
2. Double-tap to open the Settings menu.
3. Tap Device.
4. Double-tap the screen to expand the options.
5. Tap Manage Accessibility.
6. Double-tap the screen to open the Accessibility menu.
7. Tap Off next to the Screen Reader setting. Double tap to confirm.

8. Tap Continue, and double-tap the screen.

Note: *When the Screen Reader is turned off, Explore by Touch is automatically turned off.*

Built-In Apps for Fire Phone

The Amazon Fire Phone comes loaded with a variety of productive and functional apps built right into the phone. Among them are items allowing you to integrate your email accounts, browse the Internet, read eBooks, watch videos and listen to music.

Swiping up from the bottom area of your home screen will reveal all of your apps. Here's a look at some of the major apps already included on the Fire Phone.

Messaging – Use this app for sending text or SMS messages to others.

Email – With the built-in email app you can add your various accounts such as AOL, Gmail, Yahoo! and Hotmail. Use the app to browse and send your emails.

Calendar – With the special calendar app on your phone you can store important dates and set up notifications for future reminders.

Maps – Use this app to find directions and gain 3D zoom perspective of various landmarks on the map. Use the "peek" feature on the map to reveal more detailed places of interest.

Weather – Check your latest forecast based on your particular region and set other regions to check on at your leisure. The app will provide day-by-day and hourly forecasts for your current and selected regions.

Silk Browser – The Silk browser is the standard web browser for the Amazon devices including the Kindle Fire HD or HDX tablets and the new Fire Phone. There are other apps available for different browsers should you prefer to use a different one for your "web surfing" needs.

Clock – Check on the latest time, or set up various cities around the world so you can stay in the know with what the current time is there.

Calculator – Need to perform some math on the go? Use the built-in calculator to do various mathematic equations on the fly!

Books – Amazon was founded on the idea of books and reading. It carried over into their first-ever Kindle eBook readers, and their app is now

Amazon Instant Video – Amazon's exclusive online streaming video service. Prime members have access to thousands of free titles. There are also plenty of movies and TV show episodes that can be purchased to own, or rented for limited-time viewing.

Amazon Music – Just like Apple has iTunes music for their customers, Amazon also has a vast online library of digital music to select from. There are thousands of free songs and albums available for download or streaming as part of the new Prime music service. In addition, customers can purchase digital songs and albums that can be stored in their Amazon cloud or on their device.

Audiobooks – Prefer to consume your books on the go with audio playback? The built-in audiobooks app on the Fire Phone is perfect for this and will work with the many titles available through Amazon and Audible.com.

Shop – Your instant portal to shopping for thousands of products across multiple categories awaits! This is perfect for finding items through Amazon.com to order and take advantage of the Prime membership's shipping bonuses including free two-day shipping on many items, and $3.99 one-day shipping options as well.

Amazon Appstore – The Amazon Appstore is your hub for finding and downloading, or purchasing, apps, games and other content for your Fire Phone.

Games – This icon will bring up any games that were included with your Fire Phone. Some of these may include Angry Birds Free, Flow Free, Monkey Buddy, Pipes with Bridges, Planet Puzzles, and TETRIS. You can tap on any of the apps listed to download it to your phone, or tap on the store shopping cart icon in the upper right hand corner of the screen to search for other games you might enjoy.

Using Phone

While Jeff Bezos joked during the Fire Phone's unveiling that he had not made a phone call from his phone in a long time, rest assured, the Fire Phone can certainly be used to make and receive actual phone calls.

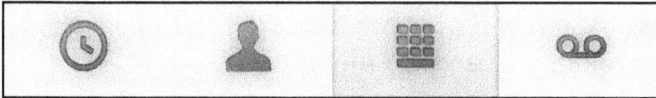

Making a Call

To make a call on the Fire Phone, do the following:

1. Tap the Phone icon in the Home screen carousel or app grid.
2. From the bottom of the screen choose History, Contacts, Keypad, or Voicemail.
3. Make phone call to Contact or from History, dial the phone number and push "Call," or check Voicemail.

Alternately, you can also make a call using Voice Dial. Complete the following steps for voice dialing:

1. Press and hold the Home button.
2. Say "call" or "dial."
3. Say the name of a contact or dictate a phone number. Be sure to "at home," "work," or "mobile" for your contact depending on where you want to call them.

Receiving a Call

There are several things you can do when you receive a phone call. The following sections explain each option.

To answer a call, complete the following steps:

1. If your phone is unlocked, tap Answer to answer a call.
2. If your phone is locked, swipe up from the green arrow at the bottom of the screen to answer. Alternatively, access the right panel and tap Answer.

To silence a call, complete the following steps:

1. Press either volume button.
2. The ringtone immediately silences.
   NOTE: *You can still answer a silenced call, until it goes to voicemail.*

To decline a call, complete the following steps:

1. If your phone is unlocked simply tap Decline.
2. If your phone is locked, press the Power button once. Alternatively, access the right panel and tap Decline.

To respond with a text instead of answering a call, complete the following:

1. Access the right panel
2. Select an available Reply-with-Text message.
   Tip: *To modify the Reply-with-Text options selecting Settings, Phone, and Edit Reply-with-Text messages. From there, simply tap on a message to edit it.*

Sending a Text

To send a text complete the following steps:

1. Click the Messaging app icon from the app grid screen or the carousel.
2. From the main screen of the Messaging app, tap the + icon at the top right of the screen.
3. Enter a phone number or name of your Contact in the To: field or tap the Add Contact icon to search your contact list.
4. Type your message and then tap Send.

Adding a Contact

Your contacts are an important part of your phone. You can access them to make calls, send text messages, send emails, share photos, and a variety of other things. To add a contact to your Fire Phone do the following:

1. Tap the Contacts app from the Carousel or the App Grid view.
2. Tap the + icon.
3. Enter the contact's details. You can tap on the "Edit image" near the top of the contact add screen to either take a photo of the person or add one from your camera roll.
4. Tap the checkmark at the top right of the screen to save the contact. You'll be asked if you want to only store the contact on your phone or also on Amazon. Once you choose one or the other, your contact will be saved for future use.

To delete a stored contact, you can simply tap on the Contacts icon, and then tap on the specific contact you want to delete. You will see that contact's info on the screen, with icons at the bottom of the screen to delete (trash can) or edit the contact (pencil). Tap the trashcan to delete.

Installing & Deleting Apps

There are many great apps for the Fire Phone, and they are simple to install. You can also easily delete an app if you download one you decide you do not like. The sections below cover how to add and delete apps.

Installing an App

The simplest way to install an app is right on your Fire Phone. Look for and tap on the Appstore icon (an orange circle with a hand and finger tapping).

This will bring up the Amazon Appstore on your Fire Phone. You can now browse through the thousands of available apps, or use the "Search" magnifying glass in the upper right hand corner of the screen to look for particular apps. (Keep in mind some popular apps may not be available for the Fire Phone yet and some of the apps at Amazon's Appstore may not be compatible with the device.)

When you've found an app you want to install, tap on it. You should get a screen for the app, with a description, screenshots from the app and customer ratings.

If the app is free you'll see an orange "FREE" button. Tap on that button and then the green "Get App" button to install the app on your device.

Once the install has completed, tap on "Open" to launch the app on your device.

Note: At the top of your apps screen you'll see the "Cloud" and "Device" tabs. Any apps you purchase or get for free at Amazon Appstore will also appear in the "Cloud" area on your phone. App icons on the Cloud area with check marks are ones that you have installed on your Fire Phone. App icons without check marks are ones you have previously purchased or downloaded from Amazon Appstore. You can tap on any app icon that doesn't have a check mark and it will download to your phone. The app should now appear in the "Device" area on your phone.

Uninstalling or Deleting an App

The easiest way to uninstall or delete an app that you no longer want on your device is by pressing down on the app icon on your Fire Phone screen. A menu will pop up with "Remove from Device." Tap on this and confirm you want to remove the app from your device on the next screen.

You can also go to the Fire Phone Settings and then tap on "Applications & Parental Controls." Tap on "Manage Applications." Scroll down the various apps you have installed to find the one you'd like to remove. Tap on that app and you will have options to manage the app including an "Uninstall" button you can tap on to remove the app from your device.

Note: *Keep in mind this will only remove the app from your phone. If you purchased the app at Amazon's app store, the app will still be associated with your Amazon account so you can download and install it again when you want to.*

Playing Music and Movies

You can play music and watch movies or videos on your Fire Phone. The following sections describe how to play music and watch movies and videos.

To play music you have on your phone, do the following:

1. From the Home screen, open the Left Panel.
2. Tap Music.
3. Choose an album from your Music Library.
4. Choose a song from the album.
5. Use the playback controls to play song.

The following explains how to stream TV shows or movies from Prime Instant Videos either for free or that you have purchased.

To watch Amazon Instant Videos or TV shows you've purchased or rented:

1. From the Home screen, open the Left Panel.
2. Toggle the Video Library to Cloud.
3. Tap Your Video Library to see all videos you have either purchased or rented.
4. Select the Movies or TV tab.
5. To stream a video, tap the title of the video you want to watch, and then tap the Watch Now or Resume.
6. To download your video instead of stream, tap the Download button.

To watch Prime Instant Video movies or TV shows:

1. Access the Videos Home screen on Amazon Prime Instant Video.
2. Find a movie or TV show.
3. Simply tap the video artwork to start streaming the movie or TV show from Prime Instant Video.

Reading eBooks

Of course a smartphone from Amazon is going to make it incredibly easy to read eBooks. There are two convenient ways to access your eBooks on the Fire Phone.

First, you can simply scroll through your Carousel and see your most recently purchased eBooks from Amazon. To read one, simply click on the book, it will download, and you can begin reading.

The second way to read eBooks on the Fire Phone is to click on the Kindle app from the apps screen. Once you are inside the app, you will see a grid of all your Kindle eBooks. You can choose the one you would like to read, it will download, and you can begin reading.

The screen on the Fire Phone is excellent for reading, which is no big surprise considering it is from Amazon.

Using Camera(s)

One huge feature of the Fire Phone is that Amazon offers unlimited photo storage in the Cloud. No other smartphone has this feature included, and it is a big plus to have a free backup of all your pictures.

The Fire Phone has a dedicated button on the left-hand side under the volume buttons. Simply push this button for quick access to the camera function. If you would rather, you can also click the camera icon from the carousel or app grid.

To use the camera to take a picture, complete the following steps:

1. Push the Camera button on the left side of the phone, or choose the camera icon from the carousel or app grid.
2. Point the phone and center the picture.
3. Tap the Shutter icon on the bottom center of the screen.

Tip: *Press the volume buttons to zoom in or out.*

To take a picture using the camera's front facing camera (i.e., a selfie), complete the following steps.

1. Push the Camera button on the left side of the phone, or choose the camera icon from the carousel or app grid.
2. Press the Front Facing-Rear Facing icon at the top center of the screen.
   Tip: *This icon looks like circular arrows.*
3. Center the picture.
4. Tap the Shutter icon on the bottom center of the screen.

To take a video complete the following steps:

1. Push the Camera button on the left side of the phone, or choose the camera icon from the carousel or app grid.
2. Tap the Camera/Video icon on the bottom right of the screen to switch to Video mode.
3. Tap the record button in the bottom center of the screen to begin recording a video. Tap record button again to stop recording.

Tip: *Press the volume buttons to zoom in or out during a recording.*

To change the camera settings, complete the following steps:

1. Push the Camera button on the left side of the phone, or choose the camera icon from the carousel or app grid.
2. Tap the Settings icon on the top left of the screen.
3. Choose to turn HDR on to capture multiple images at different exposures. Three images will be combined to create a single photo.

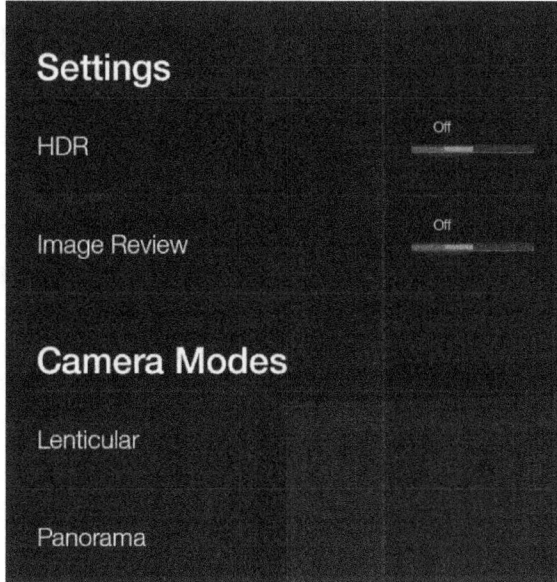

Tip: *You can also choose Lenticular or Panorama Mode in Camera Settings (shown in image above).*

Lenticular Mode:

Lenticular mode blends three images together creating a "moving" image. The Fire Phone's Photos app converts lenticular images to animated .gif files.

Panorama Mode:

Panorama uses multiple images to create a horizontal or vertical image. This camera feature works best outdoors.

Editing Photos

To edit your photos complete the following steps:

1. From the left panel on the Home screen, tap Photos, or, tap the Photos icon from the carousel or app grid.
2. Choose All, Videos, Camera Roll, Device, or Cloud Drive.
3. Tap the photo to show the toolbar at the bottom of the screen, and tap the Edit icon to open the Photo Editor.
4. Choose and editing category and tool from the bottom of the screen.
5. After editing, tap Apply to save changes.
6. When you are finished using the Photo Editor, tap Done.

Amazon Prime Features

Fire Phone owners will enjoy the Amazon Prime benefits with their phone as well as with their computers, Kindle Fire tablets, and Fire TV.

Prime Instant Video

Prime Instant Video, is a media streaming service that is brought to you by Amazon.com. This exciting service is part of Amazon Prime, which is available for $99 per year.

The downfall of Prime Instant Video is that its media streaming services include popular movies, but not always the newest releases. You can dip into the archives to find plenty of your favorites, but if you are looking for the movies that are new to DVD each Tuesday, this service lacks that up to date availability. It does, however, bring you past television episodes and kid's programming at no extra charge, but will not provide you with the previous day's television programming for free.

Using Prime Instant Video, you can watch all of their offerings as often as you like, even if you want to watch the same episode from the first season of *Downton Abbey* for the thirteenth time. Also, the service includes several TV series that are only available on Amazon, and HBO is now featured as part of this service. Plus, if you are going on an airplane or car trip with no wireless service, many of the videos are available for you to download directly to your Fire Phone.

Prime Music

With Prime Music you can enjoy unlimited streaming and downloading for more than a million songs from today's most popular artists like Daft Punk, P!nk, The Lumineers, Madonna, Bruce Springsteen, and more. The new service also includes hundreds of playlists developed by music experts for your listening enjoyment.

Using the Prime Music Feature

Prime Music allows Prime members to enjoy free unlimited streaming of over a million different songs and hundreds of different playlists on their compatible devices, including the Fire Phone. It also offers the ability to create playlists and purchase songs or albums on the go from Amazon's vast digital music store using your phone.

To use the Prime Music feature on your Fire Phone:

1. Tap on your Home button.
2. Swipe up from the bottom of the screen (where the four icons are that are directly under the home carousel.)
3. At the top of the screen make sure to tap on the "Cloud" option (rather than Device).
4. Tap on the Prime "Music" app (the icon is orange with Prime written in white and several white bar shapes on it).
5. If it is your first time with the Prime Music app, the app will begin to download and install onto your device.
6. You can now go to your apps on your Fire Phone and tap on the Amazon Prime music app to open it.

With the Prime Music app you can browse through various songs, albums and playlists. At the top of the main screen, you can also choose to refine by music genre, or perform a search to see if certain artists or albums are available to listen to on the app. As you type into the search you'll get automatic suggestions based on your typing.

A search will return various results, some of which may say "+Add" to the right of them. Clicking on the "+Add" will add those songs, albums or playlists to your Prime Music Library so you can listen to them for free. If you are looking at the songs in particular, you can also tap on the small play button over the song art to listen to a preview of that song.

Some songs or albums you find will have prices listed to the right of them indicating that you can buy those particular songs or albums. Tap on the price to bring up more info and options to purchase.

Adjusting Prime Music Settings

You can also tap on the music settings menu while you're on the Prime Music app. Menu is the 3 line icon located in the upper left-hand corner of the Prime Music screen. Tapping on this brings up a variety of options to use with your music, including a spot to enter an Amazon gift card or promotional claim code, as well as preferences for your downloading and streaming of music.

Note: *While using the Prime Music app you can also swipe from the left side of the screen to bring up a full menu of options including options to go back to the Prime Music browse/search screen. You can also choose to view your Recently Played, Recently Added and Recently Downloaded music as well as all your playlists, songs, artists, genres and more.*

If you use the Prime Music app but don't have a Prime membership, you will still be able to browse the selections of music available. You can preview and make purchases of any songs or albums you want to add to your cloud or device.

Kindle Owners Lending Library

Amazon Prime members are able to download a great book for free from the Kindle Owners' Lending Library one time each month, and using this service you have no waiting and no due dates like at a regular library. You can choose from an unbelievable number of titles, including all seven Harry Potter books, along with other current and former best-selling books.

Fire Phone Tips & Tricks

As with any device, there are plenty of tips and tricks to help you get the most out of using it. I have figured out some of the best tips and tricks for the Fire Phone and shared them below.

How to Manage Your Fire Phone Online

You may run into some instances where you are without your phone, but have access to a laptop, personal computer, or tablet. Or you might need to find or protect the phone. For example, you might misplace your phone around the home or office, or lose it while out in public. Luckily, with remote management on Amazon's site you can perform a variety of functions to locate or protect your device.

Hello. Sign in   Try
Your Account ▾   **Prime** ▾   🛒 0

Sign in

New customer? Start here.

Your Account

Your Orders

Your Wish List

Your Recommendations

Your Subscribe & Save Items

Your Prime Membership

Manage Your Content and Devices
Formerly "Manage your Kindle"

Your Prime Music

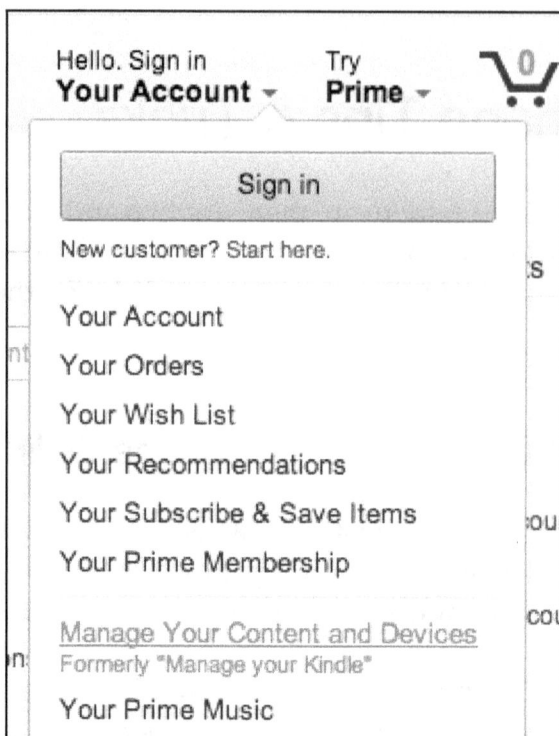

Here is how to manage your device online:

1. Log on to your Amazon.com account.
2. Scroll over to where it shows your Name and the "Your Account" drop-down menu.
3. Click or tap on "Mange Your Content and Devices."
4. You should see several large tabs on the next screen. Click or tap on the "Your Devices" tab.
5. You will now see a screen with all of your devices you have associated with Amazon or Kindle. Tap on the image that shows your Fire Phone.
6. Down below the images, you should see the name of your Phone and a "Device Actions" menu. Click on this menu to reveal a variety of helpful options.

Among the tasks you can perform online for your phone is "Find My Device" if you misplace your phone.

Note: *For this to work, you'll need to have Location services enabled on your Fire Phone as well as the "Find My Device" option enabled in your settings.*

Other options you have online include the ability to do a remote lock of your phone or a remote factory reset of your phone should it become lost or stolen.

You can also use the online management area for your phone to set a remote alarm for your device, manage voice recordings, and manage the Firefly history that Amazon stores of your various finds in images or audio.

How to Block Phone Calls

The Fire Phone did not include any sort of feature to block calls when it was first release. However, there are a few ways to get around this if you need to block calls on your smartphone.

- The first way to consider blocking calls is through the wireless service provider. At the time of this publication, AT&T is the exclusive provider for the Fire Phone. You may be able to add call blocking as a feature to your current monthly plan, so you can check with AT&T and your current account to block calls.
- A second way to block unwanted calls is by downloading and installing an app from the Amazon Appstore onto your Fire Phone. A search for "call blocker" will reveal a number of these apps including Blacklist Ultimate and Extreme Call Blocker.
- Another method for blocking calls is by using the free Google Voice service. If you have a free Google account (with Gmail) you can use the Google Voice service that gives you a free phone number. Using this service you can

---

forward your phone calls through that and use its call-blocking feature.

How to Use Fire Phone Without Service

Amazon sells its Fire Phone with AT&T service for either $199 or $299 at the time of this writing depending on if you buy the 32GB or the 64GB model. However, there is also an option to buy the Fire Phone without Service for $649 or $749 for 32GM or 64GB respectively. If you buy it without service, what are your options?

It seems that there are just a few options for the Fire Phone without signing up for a multi-year contract for wireless service. Right now, it can be used as an AT&T Go Phone where you pay as you go either with a prepaid plan or prepaid minutes depending on what plan you choose.

If you did not buy a wireless service plan for your phone, you will still have access to use it via a Wi-Fi connection. During setup, you can "Skip" the SIM card error screen to continue setting up and use your Fire Phone. With Wi-Fi you should be able to access websites and other features the phone offers.

The Skype app is currently available at the Amazon Appstore and can be installed on the Fire Phone. Many individuals use this service for making online video and voice calls, so this may be an option to use with your Wi-Fi for making calls.

You can also try Google Voice VoIP calls depending on your Wi-Fi connection. There are other VoIP apps you can use as well. One that is available via the Amazon Appstore is the "Free Calls with magicApp by magicJack" app. With this app you can sign up for a free account and phone number to use over Wi-Fi with some restrictions on calls.

How to Transfer Photos and Video to the Amazon Cloud from Old Phone

If you have photos and video on your Android or iPhone that you would like to transfer to the Amazon Cloud, you can do that. The following sections explain hot to transfer photos and video from your old phone to the Amazon Cloud.

How to Transfer iPhone Photos and Video to Amazon Cloud

If you have an iPhone, complete the following steps to transfer your photos and videos to the Amazon Cloud.

1. Download the Amazon Cloud Drive Photos app on your iPhone from the Apple App Store.
2. Sign in to your Amazon account if prompted.
3. Tap OK when asked to "Auto-Save." This will automatically save all of your photos and all videos less than 20 minutes long to the cloud.

Once you have uploaded the photos and videos to the Amazon Cloud, you can open the Photos app on your Fire phone to find your photos and videos.

Note: *Fire Phone comes with free unlimited cloud storage for photos taken with Fire Phone. You also get 5GB of free cloud storage for your other photos and videos.*

How to Transfer Android Photos and Video to Amazon Cloud

If you have an Android, complete the following steps to transfer your photos and videos to the Amazon Cloud.

1. Download the Amazon Cloud Drive Photos app on your Android phone from the Google Play store.
2. Sign in with your Amazon account if prompted.
3. Tap OK when asked to "Auto-Save." This will automatically save all of your photos and all videos less than 20 minutes long to the cloud.

Once you have uploaded the photos and videos, you can open the Photos app on your Fire phone to find them.

Note: *Fire Phone comes with free unlimited cloud storage for photos taken with Fire phone. You also get 5GB of free cloud storage for your other photos and videos.*

How to Backup Fire Phone to the Amazon Cloud

You can make sure all the information on your Fire Phone is backed up to the Amazon Cloud. That way you will not lose your pictures or videos if something happens to your device.

1. Take the following steps to backup your Fire Phone to the Amazon Cloud.
2. Go to Settings, tap Device and choose Enable/Disable auto backups
3. Turn Device Backup "On."
4. Go to Settings and tap Applications & Parental Controls.
5. Choose Configure Amazon application settings, tap Photos and turn Auto-Save "On."

How to Get More Storage

The current versions of the Fire Phone include either 32 GB or 64 GB models for on board storage. Amazon Cloud provides even more storage (including unlimited photo storage for photos taken with the Fire Phone), but some users may want another option. The best option right now is to sign up for a Free Dropbox account and download the Dropbox app from the Amazon Appstore.

Dropbox is an online cloud storage service for storing all your files, big and small. You can then access the files through the mobile app or online through the Dropbox.com website by signing into your account, meaning you'll have access to your files wherever you go and with pretty much whatever device you're using. With the option to upgrade for even more storage, this is a no-brainer for an extra storage app to install.

Later in this guide, there is also the SanDisk Wireless Media drive that is included in the recommended Fire Phone accessories. With the 64GB drive you can store up to 64GB of movies, music, videos and photos on the small pocket-sized device and then access that content from your Fire Phone on the go.

How to Take a Screen Shot

To take a screenshot complete the following steps:

1. Press and hold down the "volume down" and Power buttons at the same time on the screen you want to capture.
2. The screenshot is added to your Photos under Device.

How to Customize Lock Screen

Amazon's lock screen is a really fun part of the phone. To customize the Lock Screen, do the following:

1. Access the Quick Actions Screen by swiping down from the top of the phone.
2. Tap Settings.
3. Tap Lock Screen.
4. Select a lock screen scene.

To view available Lock Screen scenes, swipe down. You can also Your Photo to select a personal photo to display instead of one of the available scenes.

Note: *Each day, your phone will rotate through all of the available lock screen scenes. If you would like to turn this feature off, tap Off next to Rotate Scene Daily at the top of the screen or select a specific scene.*

How to Use Miracast Display Mirroring

The Miracast feature on your Fire Phone will allow you to cast or "mirror" your phone's display to a larger screen, such as a compatible streaming device, HDTV or monitor. To use screen casting from the Fire Phone to another device you'll need to make sure the other device you are casting to also has the "Miracast" feature built in.

◄ Display Mirroring

Display mirroring allows you to duplicate your Fire display and audio onto another screen, such as your TV. Please make sure the device you want to mirror your Fire display on is turned on and discoverable, then select it from the list below.

Need help mirroring your display?

Select a Device

Fire TV

One device that already offers you the ability to cast the phone's screen to your HDTV is the Amazon Fire TV. Here are instructions for how to cast your Fire Phone display onto an HDTV monitor through the Fire TV.

1. Go to Settings on the Fire Phone.
2. Tap on "Display."
3. Tap on "Share your screen via Miracast."
4. A "Display Mirroring" screen will come up next on your Fire Phone. You should now see any devices that will work with the Miracast screen-casting feature listed on this screen. You may also see that the Fire Phone is "Searching for devices..." at the bottom of the screen, indicating it is trying to find related devices you can cast to.
5. Wait for several moments if you don't see your compatible device listed on your Fire Phone. If the device still hasn't shown up after a bit of time, make sure the device you are trying to use Miracast with is turned on and discoverable.
6. For the Fire TV, if the Fire Phone connects to it you will first see a "Display Mirroring is starting soon" screen on your television.
7. Wait for several moments and you'll see a box pop up on your TV screen called "Invitation to connect."

8. Use your Fire TV remote to choose "Accept." This will start the screen casting from your Fire Phone to the Fire TV after several moments.

With this feature you'll now be able to see a mirror image of your phone's display on your TV screen as you use the phone. The display will change on your TV screen as you use the phone. You can use this option for everything from watching Amazon Prime Instant videos or YouTube videos, to checking emails, navigating the Internet, and many other tasks on your Fire Phone.

How to Use Second Screen

The Second Screen feature allows you to stream Amazon Instant Video TV shows or movies to compatible devices connected to your television. The feature can only be used with content you have purchased from Amazon Instant Video, not with free Prime videos.

When you go to your purchased content on Amazon Instant Video, you will see a small second screen icon (an up arrow and small TV screen) next to the Watch or Resume button on the particular movie or TV show you want to use the feature on.

To Use Second Screen

1. Tap on the Second Screen icon located on the "Watch" or "Resume Playback" button on your Amazon Instant Video content.
2. A small window of available devices you can use to stream the video to will pop up on your Fire Phone. Tap on the device you want to stream to (i.e. Fire TV, PlayStation 3 or 4, Samsung TV).
3. You can now watch the movie or TV show on your second screen. The Fire Phone can act as a remote control for the video to pause and resume playback, or rewind back. You can also press the home button and use other aspects of your Fire Phone while the video continues to play on your second screen.

Note: *For PS3, PS4 or other devices you may need to make sure you have the Amazon Instant Video app updated on the device and then make sure it is running to use the Second Screen feature.*

To Stop Second Screen

1. To stop the video from playing on your second screen, you'll need to go back into the Amazon Instant Video on your Fire Phone.
2. Tap on the second screen icon in the upper right-hand corner of your screen.
3. Next, you'll see a pop up menu of devices. Tap on your Fire Phone to transfer the video back to just your phone. The video will continue on your phone and stop playing on the other screen.

How to Use Visual Voicemail

Visual voicemail shows you a list of your voicemail messages. From this list you can choose which one to listen to or delete. You can see how many unheard messages you have by checking the orange badge on the Voicemail icon.

Complete the following steps to set up visual voicemail:

1. Tap the Phone app.
2. Tap the Voicemail icon.
3. Tap Call Voicemail and follow the instructions to set up a voicemail password or PIN and select or record a voicemail greeting.
4. When the call ends, type in your voicemail password or PIN.
5. Tap then Continue and finish setting up visual voicemail.

To listen to a voicemail message simply:

1. Tap the message you want to hear.
2. Press Play.
3. When you are finished, you can tap the Delete icon or press and hold down the message and tap "Delete Voicemail."

How to Set a Lock Screen PIN or Password

A lock screen appears when you wake your phone. If you would like to set a PIN or password for your phone, complete the following steps:

1. Access the Quick Actions screen by swiping down from the top of your phone.
2. Tap Settings.
3. Tap Lock Screen, and then tap Set a password or PIN.

4. Select whether you want to use a 4-digit PIN, or a Password that contains a combination of letters, numbers, and special characters.
5. Enter the PIN or password, and tap Continue.
6. Confirm the PIN or password, and tap OK.

Note: *You can also select the amount of time your phone stays on before it automatically locks under the Lock Screen settings. You can also enable Notifications to display on the Lock Screen from the Lock Screen settings.*

How to Use Bluetooth

Some of the items you can connect to and pair with the Fire Phone using Bluetooth include wireless speakers, headsets, and other devices such as tablets or laptop computers.

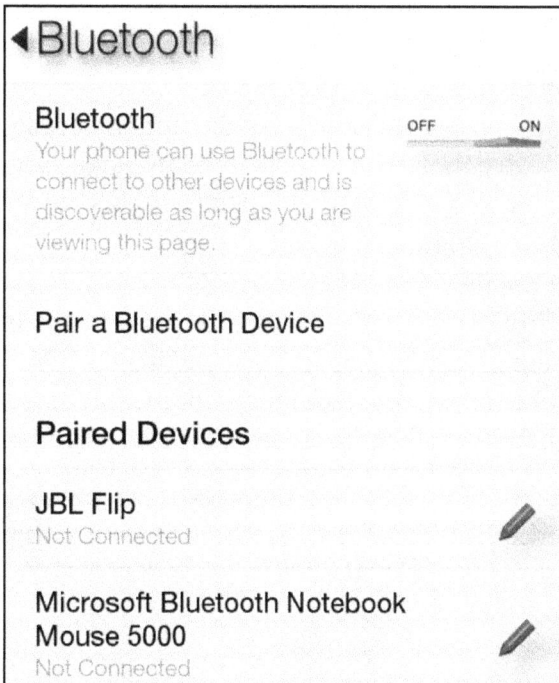

◄ Bluetooth

Bluetooth   OFF   ON
Your phone can use Bluetooth to
connect to other devices and is
discoverable as long as you are
viewing this page.

Pair a Bluetooth Device

Paired Devices

JBL Flip
Not Connected

Microsoft Bluetooth Notebook
Mouse 5000
Not Connected

As examples, you can use Bluetooth to connect to a laptop such as a MacBook Pro or other laptop and then send files back and forth using the connection. Or you could pair a compatible wireless Bluetooth speaker with the Fire Phone to listen to audio over the speaker. Even compatible Bluetooth mice and keyboards can be paired to and used with your Fire Phone.

Some models of vehicles may also have Bluetooth integrated into their features, allowing you to make and receive phone calls over your car stereo system. It is important to refer to any included documentation with your other device for proper setup and connection with Fire Phone.

To Pair or Connect a Compatible Bluetooth Device to Fire Phone:

1. On the Fire Phone, go to Settings.
2. Tap on "Wi-Fi & Networks."
3. Tap on "Pair Bluetooth devices."
4. Make sure the Bluetooth slider is switched to the "ON" side.
5. Tap on "Pair a Bluetooth Device."
6. Your Fire Phone will start scanning for any available Bluetooth connections. You'll need to make sure the device you want to pair with or connect to is also powered on and that it has Bluetooth turned on.
7. When the item you want to pair or connect to shows up, tap on it. The Fire Phone will begin to pair with the device. (This may require tapping on an "OK" or "Accept" window on the other device and the Fire Phone, or making sure a number code matches up with the two devices on their screens first.)

Once the items are paired or connected, the device should now be saved in the Bluetooth settings area of your Fire Phone for future connections.

How to Use Fire Phone as a Wi-Fi Hotspot

The Fire Phone can act as a Wi-Fi Hotspot, allowing up to five different devices to connect to it wirelessly and use Wi-Fi. You will need to have a data plan associated with your wireless service in order to use the feature.

Here is how to set it up:

1. Go to Settings and then tap "Wi-Fi & Networks."
2. Tap on "Set up a Wi-Fi hotspot."
3. Switch the Wi-Fi Hotspot slider to "ON."
4. You can also tap on "Configure Hotspot" to set up the name of your Fire Phone hotspot and security for it. You can leave the network open, meaning any nearby device might be able to connect. If you'd prefer to make your network more secure, you can create a log-in password and even hide your network name from devices.

To turn off your Fire Phone as a Wi-Fi hotspot, go to Settings, and then tap "Wi-Fi & Networks." Tap on "Set up a Wi-Fi Hotspot," and then switch the Wi-Fi Hotspot slider to "OFF."

How to Watch Video on Fire Phone

The most basic ways to watch video on the Fire Phone involve the use of the Amazon Instant Video service. If you're a Prime member, you've got access to a large catalog of free movies and TV shows which you can instantly stream and watch. There are also many more movies and shows you can opt to purchase and download whether you are a Prime member or not.

Other options for watching video on the Fire Phone include the use of the Silk browser or another web browser, such as Opera Mini. You can navigate to popular video sites such as YouTube to begin watching their millions of available videos.

Apps for the Fire Phone also exist that will complement any monthly services you might currently use (or sign up to use) for watching content. These include Netflix, Hulu Plus, Showtime Anytime and HBO Go. There are also other video services such as Vimeo or Vevo available in the form of apps, with the Vimeo site allowing for downloading of select videos.

How to Watch Videos Offline Later

Watching videos on the go when you may not have access to good Wi-Fi or wireless service is something many Fire Phone owners will want to do. At the time of publication, there are currently a few options for doing this.

There are a variety of YouTube downloader apps available at Amazon Appstore. The best versions of these will allow you to download YouTube videos when you have a Wi-Fi or wireless service connection available. You can then watch your downloaded videos at another time. The Fire Phone has a built-in video player that will allow you to watch most video file types you have downloaded to the phone offline at a later time.

Some of the downloader apps will function offline to allow you to watch your downloaded content, while others may require use of a free file-opening app like File Expert HD, which is available at Amazon Appstore.

Another option is to download videos to a laptop or personal computer and transfer them directly to your Fire Phone using the micro USB cable. You can also opt to purchase and use a wireless media drive for more offline storage of content. One great drive is the SanDisk Wireless Media Drive, which is also described in the accessories section of this guide.

There are two different storage sizes available for the device, with the biggest version able to hold up to 64GB of music, videos and photos. You can connect to the drive wirelessly on the go and stream videos from it to your Fire Phone.

How to Sideload Third Party Apps

Sideloading is a way of loading third party Android apps onto devices using special app files known as ".apk" files. At this point, many of the Android apps that are not available at Amazon's app store may not be compatible with the Fire Phone.

However, some third party apps may work with the device. It is important to note that this practice of sideloading and installing other apps can bring its share of issues, so use this tip with caution and read the warnings carefully on the Fire Phone.

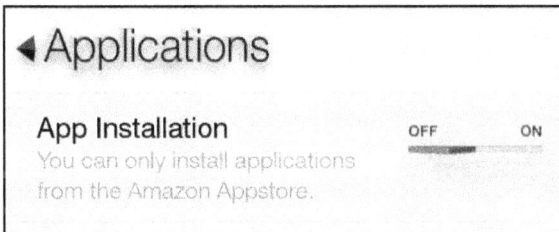

◄Applications

App Installation   OFF   ON

You can only install applications
from the Amazon Appstore.

Here is how to sideload those third party apps that might work on the Fire Phone:

1. Go to Settings and tap on "Applications & Parental Controls" on the Fire Phone.
2. On the menu options, tap on "Allow non-Amazon app installation."
3. Ensure the first option, "App Installation," on the next screen is clicked to "OFF."
4. You'll get a pop-up warning from Amazon about allowing third party apps on the Fire Phone. After reading the warning message, click "OK" to allow installation of third party apps.
5. To open and install an APK file, you can use File Expert HD, a free app available at the Amazon app store. Install the free app on your phone.
6. Next, download or save the .apk file you want to use on your Fire Phone on a computer.
7. Connect your Fire Phone to the computer using the micro-USB cable.
8. Move the .apk file from your computer onto your Fire Phone. Make sure you place the .apk file into a folder you'll remember on the Fire Phone.
9. Open File Expert HD on your Fire Phone.
10. Navigate to the "Folders" tab and tap on it.
11. Find the folder that you moved the .apk file into and tap on it.
12. Find the .apk file and tap on it. You'll now go through several screens to install the .apk file and Android app on your phone.
13. Tap on "Open" to open the sideloaded app on your phone. You will also be able to see the new app icon on the screen with all your other Fire phone apps.

Note: *Keep in mind, that a select number of .apk files may currently work with the Fire Phone, but you should already own the .apk file before using it. It's not recommended that you download the apk files from third party sources online, unless you fully trust the site that the download is presented on. Some third party sites may also try to trick you into downloading extra files, or .apk files that are merely apps for their download software, so tread cautiously when doing this.*

How to Set Up Parental Controls

If you are a parent or have young children who use your phone, you will probably want to set up parental controls for the Fire Phone. These controls will help you restrict certain activities and content on the Fire Phone.

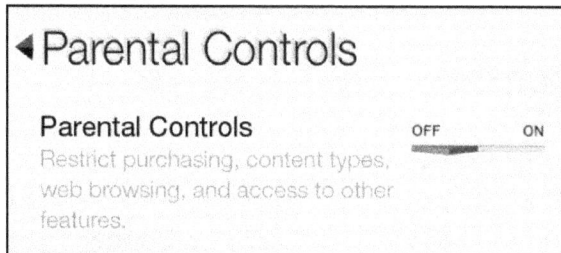

◀ Parental Controls

Parental Controls   OFF   ON
Restrict purchasing, content types,
web browsing, and access to other
features.

To set them up, do the following:

Access the Quick Actions panel by swiping down or swiveling phone.

1. Tap Settings.
2. Tap Applications & Parental Controls
3. Tap Enable Parental Controls.
4. Move parental controls switch to ON.
5. Enter a password, confirm password, and tap Submit.

Once the password is set, you can restrict access to one or more of the following options depending on your parental control needs and wishes:

- Web Browsing
- Email and Calendars
- Social network sharing
- Camera
- The ability to purchase from the content stores on your device (for example, the Amazon Appstore)
- The ability to play movies and TV shows from Amazon Instant Video
- Specific content types (for example, Music, Video, Books, or Apps)
- Wireless and mobile network connectivity
- Location-Based Services

Note: *If you would like to modify or disable your parental controls settings, tap Disable Parental Controls in the Settings menu. You can either update your restrictions or turn parental controls OFF.*

How to Conserve Battery Power

The Fire Phone offers so many unique options, and these can drain the battery life quickly. However, there are a few tips for conserving battery power for your Fire Phone.

Turn off Dynamic Perspective

One way to save battery power on the Fire Phone is to turn off Dynamic Perspective. To do this, complete the following:

1. Swipe down from the top of the phone to open the Quick Actions panel.
2. Tap Settings.
3. Tap Display Settings.
4. Tap Configure Low Motion Settings.

5. Move the Dynamic Perspective toggle switch to "Off" to turn off Dynamic Perspective.

Note: *You can also adjust other battery draining display features here like Tilt, Peek, Swivel, and Auto-Scroll. You can also put the phone in Low Motion Mode to disable everything at once.*

Adjust Screen Brightness

Adjusting the screen brightness can also conserve battery power. To do this, simply:

1. Swipe down from the top of the phone to open the Quick Actions panel.
2. Tap Settings.
3. Tap Display Settings.
4. Tap Adjust screen brightness.

In this area, you can turn Auto Brightness off, and then adjust Display Brightness to the lowest possible setting to conserve battery power.

Change Time to Sleep

Another way to conserve batter is to Change time to sleep. Simply complete the following to adjust this feature:

1. Swipe down from the top of the phone to open the Quick Actions panel.
2. Tap Settings.
3. Tap Display Settings.
4. Tap Change time to sleep.
5. Tap 15 Seconds to conserve the most battery power.

Use Wi-Fi When Possible

Using Wi-Fi uses less battery than using mobile networks. Whenever you are in range of Wi-Fi be sure to enable it in order to conserve battery. To do this, complete the following steps:

1. Swipe down from the top of the phone to open the Quick Actions panel.
2. Tap Settings.
3. Tap Connect to Wi-Fi.
4. Tap "Scan" to scan available networks.
5. Tap the network you want to connect to, and enter password details to connect, or if it is an unprotected network, simply connect without entering a password.

Keep Software Up to Date

The latest software is likely to have the best battery power conservation available for the phone. To make sure you have the Fire Phone's latest software, do the following:

1. Swipe down from the top of the phone to open the Quick Actions panel.
2. Tap Settings.
3. Tap Device.
4. Tap Install system updates.
5. Install any new updates that are available.

Keep Fire Phone Charged

Aside from keeping your Fire Phone plugged in to the micro USB wall charger whenever you're not using it at home or the office, you might also consider investing in several helpful accessories.

There are USB chargers that will plug into the cigarette lighter spot of your vehicle and charge your device. There are also portable USB charging devices you can purchase that are small and compact, yet will charge up your device on the go. There is more information on these accessories later in this guide.

How to Factory Reset Fire Phone

You may want to factory reset your Fire Phone if you are giving it away, selling it, or for some other reason, such as troubleshooting an issue with the phone. Performing this operation will return your Fire Phone to its original state, just how it was from the factory. That means all content on the device will be erased, all accounts will be removed from the device, and all settings will be returned to their original state.

◀Factory Reset

Resetting your Fire to factory defaults will remove all your personal information, Amazon account information, downloaded content, and applications.

Reset

Reset to Factory Defaults

Are you sure you want to reset to factory defaults? All data will be removed.

OK

Cancel

To factory reset your phone, do the following:

1. Swipe down from the top of the phone to open the Quick Actions panel.
2. Tap Settings.
3. Tap Device.
4. Tap Factory reset.
5. Tap Reset.

20 Great Apps for Fire Phone

While the Amazon Fire Phone comes pre-loaded with a variety of helpful apps, there are many more available through the Amazon App store online. Here is a breakdown of some of the best apps to download and use on the Fire Phone right now.

Evernote

Evernote lets users create a central area for storing and reviewing information, whatever information that might be. Evernote can be used to store reminders, to-do lists, pictures, and even voice recordings. All of this information is neatly organized and easily searchable. Notes can be synced from multiple mobile devices, tablets, and computers. All of this data can then be stored on an external memory source as a backup.

Facebook

Facebook is the world's most popular social networking website. The Facebook app brings all of the friends, likes, photos, and shares to the user's hand without requiring a browser. The app is fully optimized for mobile devices and many users prefer it to the Facebook mobile website. It can be used by anyone who has an existing Facebook account.

iHeart Radio

The iHeart Radio app is the only application that brings users live radio stations from around the world. These aren't online or satellite radio stations, but rather genuine FM radio stations that play through the radio on a daily basis. This application gives users the chance to listen to their favorite radio stations, even if they aren't in the same state anymore.

Instagram

More than 200 million users agree that Instagram has created the perfect platform and application for sharing photos. Instagram lets users capture, edit, and share all of their favorite moments. They can also upload existing photos from their phone. Users create their own profiles and have access to live feeds of friends.

Pandora

Pandora lets users listen to all of their favorite music as much as they like and for free. Users have access to a large library of existing stations that feature millions of songs from all genres. If you don't like the existing stations, you can create your own stations that plays only the songs you like. Stations can be searched according to genre, artists, or keywords.

Pinterest

Pinterest is a giant digital collection of all the things that people love. Users create a profile and then begin creating their own personal projects. Projects, ideas, and "pins" can be shared amongst users. Projects and pins are created on public boards. There are thousands of existing projects already on Pinterest, which makes it the perfect place to get some great ideas.

Skype

Skype is a mobile messaging application that has integrated voice and video calls via VoIP. Millions of people already use Skype because it allows anyone to stay in touch with the people that matter for free. It also allows calls directly to mobile phones and landlines for a small fee.

StubHub

The StubHub app lets users quickly and easily get tickets to their favorite events, concerts, and sports games. Anyone who frequents public events such as these will low using StubHub. It offers some of the best prices and the prices shown are the prices charged without any additional fees. Users can find, purchase, and get their tickets all from a mobile phone.

Twitter

Twitter is a free social networking app that lets users connect sharing short posts of 140 characters or less. This is just enough room to say all of the important things without the bells and whistles. People from all over the world use Twitter. Users can share text, photos, and even videos to their followers.

Uber

Uber makes catching a taxi a thing of the past. It's available in more than 130 different cities across 30 different countries. Users can instantly request a ride, which will then appear within a matter of minutes. There are absolutely no reservations required and no waiting time. Users can get where they need to be before they need to be there.

WhatsApp

WhatsApp is a widely popular instant messaging app available for most modern smartphones. It can use Wi-Fi, 3G, or 4G connections to instantly share messages with anyone in the user's WhatsApp contacts. WhatsApp makes it easy to send videos, images, and other multimedia messages without breaking the bank. It also has a neat group chat feature that allows multiple users to share one conversation.

Yelp

Yelp helps users find local establishments that are actually worth their time and money. Whether it's a gas station, a business, a restaurant, or a bar, there are usually a couple of reviews found on the Yelp application. After visiting an establishment, users of Yelp post their reviews to help other users know where to go and what to avoid. It's useful to just about everyone in the world who's interested in finding a great meal or the best hotel.

Zillow

Zillow is a real estate app for mobile devices meant to show users available homes for sale in a certain area, as well as their value and other important information. It's a one-of-a-kind app that is useful for anyone interested in buying a home in the future. It can also be used to locate rental homes and apartments anywhere in the United States.

Fit Brains Trainer

The Fit Brains Trainer provides games on your Fire Phone, that are not only enjoyable, but that are designed to help the brain develop and improve in a wide variety of areas, including memory, problem-solving, increased processing speed, concentration, and much more. A great way to keep developing and flexing your brainpower on the go!

Postagram Postcards

There is now an app that allows you to turn your own photo, into postcards, and send those to your loved ones from your Fire Phone. Postagram Postcards have an additional benefit to the recipient - the photo can be popped out and saved, when the card is discarded! There are a variety of free postcards to choose from, which have sponsored branding on them, or you could pay a small fee for postcards without branding.

Adobe Reader

Although the Fire Phone comes with its own built-in app for reading PDF files, it is not as versatile or flexible as Adobe. The Adobe Reader app allows you to view PDF files in various views, for example, single page or continuous, and it also allows you to easily and quickly send a PDF file as an attachment by email. Unlike the app that comes with Kindle Fire, Adobe Reader also allows you to search throughout the whole document.

Opera Mini Web Browser

While Silk browser is built into the Fire Phone, some users may want another option. Opera's app is fast and easy to use, and may be much more efficient browser than the default Silk browser.

Because of its high speed, it's particularly perfect for those who are paying for their connection per megabyte, or who are on slow connections. This app also comes with text wrapping, find in page search, customizable privacy settings, pinch-to-zoom, and tabbed browsing with long-press menus that allow you to copy and paste, as well as manage tabbed content just as you would on a computer.

Any.do

The next level in "to do" lists has arrived! Any.do is an app that helps you organize your day and your life in an easy, efficient way, on your Fire Phone. The free app allows you to get things done in an organized way, and it can sort tasks by priority, folder or date. Use the reminders option to make sure you don't forget the most important tasks, and invite friends and family to join you by syncing on specific lists and tasks.

Flipboard

This app is a great way to gather information on your Fire Phone, from a wide variety of sources, and place it into your own magazine format. It looks great, and is easy to read and browse. Flipboard is effectively your own personalized interactive electronic magazine, and you have the option of adding Facebook and Twitter into the interface. It's a user-friendly, convenient way to read the news, Twitter and Facebook feed in a condensed magazine format.

magicApp by magicJack

For unlimited free calls on your device, within the United States and Canada, and unlimited free calls between magicJacks worldwide, there is the free magicApp.

You can start making and receiving free calls with magicApp by downloading the app and logging in or signing up for a free account. You'll receive a special magicJack number, and you can use the same number on your magicJack device, and on the magicApp. This is a great download to use your phone to make calls over Wi-Fi rather than using your wireless service, although remember you'll be using a different number.

Bitdefender's Antivirus Free

This great free antivirus app uses "in the cloud" scanning technology in order to give your Fire Phone the most effective virus detection, without slowing it down or draining the battery. When on autopilot, the app automatically scans new apps for viruses, providing peace of mind, with hardly any impact on battery life and without affecting the performance of your phone.

10 Great Games for Fire Phone

Not only can you download all sorts of great apps for your Amazon Fire Phone, but you can also download free or paid mobile games to keep you entertained on the go. Some of the popular titles include Lili, Sabre's Edge, Minecraft, Cut the Rope 2, Angry Birds Go, Jetpack Joyride, and To-Fu Fury. Here is a look at some of the best free and paid games to get right now including a few exclusive titles available for the Fire Phone.

Angry Birds Go (Free)

Welcome to downhill racing with Angry Birds Go on Piggy Island! You will feel the excitement as you throw your birds at the piggies while racing downhill. There are plenty of thrills and chills with this exciting new game but you need to beware of hazardous road conditions and evil opponents who are racing right behind you.

Minecraft ($6.99)

While it may be costly, this is a vastly popular mobile game featuring various adventures and placing blocks. There are two modes to the new pocket edition game including creative mode and survival mode. As you travel through the infinite worlds of villages, mobs, caves, and other exotic locations you will be able to explore to your heart's content. This is a fast action game that is thrilling to play with fans all around the world.

Lili ($2.99)

If you like adventure on mysterious islands Lili will introduce you to a variety of beautiful environments and crazy but hilarious locals. You will be able to use your brainpower to solve challenging puzzles and thwart oppressive regimes. Lili is a non-combat type game that is fine and also allows you to sort out a few personal issues as you play.

Saber's Edge (Free)

Are you ready for high action adventure with Saber's Edge? This game is challenging for those who like to play strategic puzzle games. For this mobile game, you'll plan your defenses, chain together attack strategies, and defeat your worst foes. As a strategist you will be able to outflank, outmaneuver, and outlast your enemies to become famous as a renowned Raider. Look for treasure and adventure as you set sail across the open skies.

You'll be competing against other deadly adversaries to achieve the fabled Eye of Markesh. This is no easy task, as you will have to battle against evil Dark Mechanics and other nefarious scoundrels. You'll to be at your best when it comes to skills because there is going to be battles going on everywhere, including every port. Are you up for the challenge?

To-Fu Fury ($1.99)

This is a great action-packed mobile puzzle game from Amazon Game Studios that is exclusively for the Fire Phone. For this particular game, you will be required to win battles and solve crafty puzzles. There are deadly traps and devious bandits as you walk along your wooden surfaces while at times sliding down slippery jades. The game features a champion who might look a bit funny but he'll help you to overcome obstacles and defeat your enemies. Don't be deceived by his silly looks, as his lightning speed and fighting skills will make him the perfect companion for your quest.

Cut the Rope 2 (Free)

Cut the Rope 2 is another free mobile game available for your enjoyment from the Amazon App Store. The sequel to the popular Cut the Rope features many new challenges in the new game. You'll be guiding the main character Om Nom through new candy collecting adventures and missions to earn medals and advance farther along in the game. Along the way, he'll have the help of new friends such as Roto, Lick, Blue, Toss and Snailbrow, who will help Om Nom in his hunt for candy. If you loved the original, you'll love the challenging obstacles you'll face in this new title.

Diner Dash (Free)

Diner Dash is an entertaining app with strategy at its core. The game focuses on managing a growing diner and all of the responsibilities that come with it. Flo, the main character, quits her tedious job at a stock market company and decides to open her own diner. Gameplay involves strategic time management while performing all of the necessary tasks of a successful diner, such as seating guests and serving customers. Completing these tasks allows Flo to earn enough income to expand and improve the diner, which adds new features, responsibilities, and customers. Diner Dash is fun enough to entertain the kids, but challenging enough to keep the adults busy too.

Temple Run (Free)

Temple Run is a fast-paced, action packed running game without an ending. Players play as an adventurous treasure hunter who is now being chased by crazy monkeys who want their treasure back. The character begins running at the start of the game and only stops once he or she dies. Running requires moving the phone to dodge obstacles and turn around sharp corners. Touching the screen causes the player to jump over gaps, trees, fire, and other dangerous hurdles. There's no beating the game: It's all about getting the highest score possible, with running becoming exponentially harder as time passes. Those who have played the first version of the game can also check out Temple Run 2 for even more challenges!

Words With Friends (Free)

Words with Friends was designed as a multiplayer game and a single player can participate in up to 20 different games at one time. Gameplay can happen quickly, over a period of days, or according to certain clock limits. The rules and play style are very similar to the board game Scrabble. Players use a pool of existing letters to build words for points in a crossword fashion. It's a very popular game, so there's never a problem finding a random person to play with over the Internet. The app can also find friends from Facebook and you can invite them to play games against you.

Jetpack Joyride (Free)

Jetpack Joyride is an exciting endless running game that takes place from a side scrolling perspective. The hero of the game has just escaped a top-secret laboratory using their jetpack and needs to get away immediately. Rather than control speed and direction, players simply touch the screen to cause Barry, the main character, to use his jetpack. When touching the screen, the jetpack is activated and Barry ascends higher. If you don't tap or press the button enough he will start to fall because he is in constant motion.

Fire Phone Troubleshooting

You may have problems while using your Fire Phone. The next sections describe some common issues that may occur on your Fire Phone along with some common fixes. Hopefully they will be of help when you are having trouble. This is certainly the place to start when you have issues with your phone.

Frozen or Unresponsive Phone

If your Fire phone screen is frozen or simply unresponsive in general, the first thing to do is restart your phone. To restart your Fire phone complete the following:

1. Press and hold the Power button.
2. Select Restart from the options available.

This will often fix a frozen or unresponsive phone.

Issues with Specific Apps

If you have issues with a specific app, you can try the steps below to correct the issue.

1. From the Home screen, swivel your phone or swipe down from the top of the screen to open the Quick Actions panel
2. Select Settings and tap Manage applications.
3. Select the application you are having difficulty with. From here you can change app settings, clear data, clear cache, force stop, or uninstall the problem application.

Note: *If you clear data, you will not delete the app; however, any saved information like game scores or account information could be lost.*

Purchased Content Not Showing

If your purchased content is not showing, you need to verify that your Fire Phone is connected to the correct Amazon account.

1. Swivel your phone or swipe down from the top of the screen to open the Quick Actions panel.
2. Tap Settings, then tap My Accounts, and tap Deregister your phone.

Note: *If you see the wrong account, simply tap Deregister to confirm. After you deregister your phone, simply tap Register to register your Fire Phone to the correct Amazon account.*

Fire Phone Not Charging

If your Fire Phone is not charging, there are several things you can check to try to fix the issue.

- Make sure you are using the power adapter and micro-USB cable that came in the box with the phone.
- Make sure you are charging your phone from a power outlet instead of a USB port.
- Make sure the cord is securely connected to the Fire Phone and to the power adapter.

If you have checked all these things, and the phone still will not charge, take the following steps:

Unplug the power adapter and micro-USB cable from your Fire phone, and then connect them to your Fire phone again.

1. Insert the power adapter into a power outlet.
2. If your Fire phone does not indicate that it is charging, simply unplug the power adapter. Then, restart your phone by holding down on the Power button, and tapping Restart.
3. After restarting your Fire phone, plug the power adapter and micro-USB cable into the device and charge for at least an hour.

Forgot Fire Phone Password or PIN

If you forgot your PIN or password, there is some good news. You can actually change the password or PIN from the Manager Your Content and Devices page on your Amazon account.

1. Go to Manage Your Content and Devices, and click on Your Devices.
2. Choose your Fire Phone from the list of devices registered to your Amazon account.
3. Open the Device Actions drop-down menu, and choose Remote Lock.
4. Enter and confirm a new password or PIN, and select Lock Device.
5. From your Fire phone, tap Unlock Device.
6. Enter your lock screen password or PIN, and tap OK.

Keep in mind that some issues may arise which relate to the phone, and some may relate to the wireless service, or your Internet provider. The Mayday button and related feature on the Fire Phone is a great way to get a live tech support rep from Amazon to help you with a variety of problems.

If you bought your phone from a wireless service provider store or their online site, consider consulting those sources for more help. For additional help with your Fire Phone, you can also consult the additional resource links towards the end of this guide.

Fire Phone Accessories

The Amazon Fire Phone is a great new device on its own, but there are plenty of ways to protect it, give it a more stylish look or feel and enhance some of its best features. These great accessories for the Amazon Fire Phone will help your phone last longer, look better, and be much easier or more convenient to use.

Protective Case

A protective case will help prevent accidental damage to your phone. At the time of this publication, the SUPCASE for the Amazon Fire Phone is a great value for the money spent on it. It is a full-body, hybrid case that is hardwearing and looks great. Choose this case to protect your phone from scratches, dents and falls, ensuring that it looks great and serves you well for a long time to come. Amazon also makes their own great smartphone and tablet cases in addition to other manufacturers out there that sell for a bit more, but will certainly help protect your item and add color or style.

Screen Protector Film

The Moshi Screen Protector is a clear protective film that features a non-adhesive backing so it is easy to remove and replace without having to worry about sticky residue. The protective film helps to keep your screen free of scratches and fingerprints, and there is also a type that serves to reduce glare when using the phone in bright sunlight. There are also other brands available on the site for a lower cost, but won't quite have the same quality of the Moshi protector.

Portable USB Charger

A lipstick-sized external battery charger is designed to allow you to charge your phone on the go. It works not just with the Amazon Fire Phone, but also with almost any other USB-chargeable device. One type to consider is the Anker Astro Mini charger. There are other more expensive battery chargers out there, which provide more power on the go and are a bit larger than a deck of cards. Pick up one of these handy devices and you will never need to worry about running out of battery power again.

Speaker

If you like to listen to music via your phone, then this Bluetooth speaker is a must-have. This wireless stereo system will let you stream your favorite tunes from your phone to a powerful speaker system, so you can enjoy them the way they're meant to be heard, instead of from weaker phone speakers. Since the JBL Flip speaker relies on rechargeable batteries and Bluetooth, you don't have to worry about tangled wires or finding somewhere to plug the speakers in. Another great feature of this phone is that it can potentially take phone calls over the speaker. This feature has worked on many other smartphones when owners are using the speaker to listen to music from their smartphone and a call comes through. It may or may not work the Fire Phone.

USB Car Charger

If you don't already have a USB car charger, you should definitely add one to your shopping list. The Amazon Fire Phone is far more than just a phone - you can use it for listening to music and podcasts, reading books and watching videos. As an all-in-one entertainment system, the battery will be under a lot of stress. Why not top up the battery while you're driving? The Amazon Basics 2-port charger will easily help to charge up your smartphone and one other device on the go!

Car Mount Holder

The iOttie Car Mount Holder is designed for use with the Amazon Fire phone and other popular phone brands of smartphones. This versatile phone holder can be placed on your vehicle's dashboard with a strong suction cup. It's also movable with 360-degree rotation so that the phone screen can be comfortably read, no matter where in your car you decide to mount the holder. An easy one-touch button helps release your Fire Phone from the iOttie's grip with the simple push of your finger.

Wireless Media Drive

Gone are the days when phones were used exclusively for making calls and sending text messages. Today, we use our phones for social media, photos, storing music, videos and podcasts, and much more. While the Amazon Fire Phone has a decent amount of storage, you don't always want to keep everything on your phone. A wireless media drive can be easily hooked up to other devices and is a great way to share important files. The SanDisk Wireless Media Drive range includes drives up to 64GB in size. That generous storage is ideal for loading up entertainment in preparation for a long flight or a few days away with work.

High Quality Stereo Earbuds

When listening to your device privately, it can be tempting to just use the earbuds that come free with your device, but for audio enthusiasts, this is a mistake. Not only do the stock earbuds usually have inferior audio quality, they are likely to break fairly quickly.

For many individuals, the included earbuds work perfectly fine with the device. For others, the sound and audio quality of the earbuds included with Fire Phone may not meet your expectations. Sennheiser are a world leader when it comes to high quality audio, and their enhanced bass earbuds are perfect for any audiophile. Picking up a pair of high quality earbuds from brands like Bose or Sennheiser will ensure you've got great quality sounds and a pair of earbuds that will last a good bit of time.

Stylus Pen

The Amazon Fire Phone has a touch screen that you can use easily with your fingers, but many of the creativity apps offered for the device demand more precision than most people can get with just a finger. Purchasing a stylus pen is ideal for use with apps that involve drawing or writing, and is also useful for people with bigger hands who find that they struggle to use their fingers to operate touch-enabled apps.

Bluetooth headsets

Currently, the selection of Bluetooth headsets that work with the Amazon Fire Phone is quite limited. Rather than mail ordering a headset and hoping that it works, the best thing to do is pay a visit to your local AT&T store. They will be able to advise you on which headsets work with the phone. A Bluetooth headset is a must-have for anyone that spends a lot of time out-and-about. It frees you from the hassle of worrying about cables, and lets you talk on your phone hands-free.

Amazon Fire TV

If you already own the Fire Phone and also own another Amazon mobile device such as the Kindle Fire HD or HDX, you'll definitely want the Amazon Fire TV to connect to your HDTV. This accessory which typically sells for $99 on Amazon, allows you to stream media to your HDTV. It also offers a variety of apps that can function on the Fire TV, as well as plenty of games, and the option to store additional content in the cloud such as photos or videos. You can find additional info on Fire TV in our related guide book listed towards this end of this book.

For those that love Amazon Instant Video content, this is a must-have accessory to go with the Fire Phone, as it will allow you to perform features such as Miracast screen mirroring or Second Screen from your phone to your TV display.

There are many other accessories and add-ons out there that can make your Amazon Fire Phone even more fun to use, but the above are some of the most interesting and useful accessories. Expect more accessories to arrive as more companies develop more specific items to complement the Fire Phone.

Conclusion

Amazon has done an amazing job with its first entry into the world of smartphones. The Fire Phone is a versatile phone that offers what you want in a smartphone along with some revolutionary features like Dynamic Perspective and Firefly. After using these for a while, you will probably find that you do not want to go back to a "regular" smartphone.

The Fire Phone is a whole new world for Amazon, and if you are a fan, it seamlessly integrates with your other Amazon devices and services like the Kindle Fire HDX, Fire TV, and Amazon Instant Video. It is a must-have phone for people who love Amazon, its products, and its services.

Keep in mind that this phone is in its very first version too. No doubt the Fire Phone will continue to improve as customer comments roll in to Amazon. Be sure to keep your software up-to-date because there is no doubt that subsequent versions of the operating systems will add upgrades and new functionality to your phone.

Take a bit of time to get to know this phone, so that you can truly unleash the amazing possibilities it offers. This guide should have you well on your way to getting much more out of your phone than you previously could!

If you feel that this guide provided value and was a helpful resource, please leave an honest review to help other customers who may be looking for help with their device.

More Fire Phone Help & Resources

Amazon Fire Phone Support Page <http://amzn.to/WI42QF>

Amazon Fire Phone Customer Q&A Forum <http://amzn.to/1AjqIWy>

AT&T Wireless Support Page <http://www.att.com/esupport/main.jsp?cv=820>

AT&T Customer Q&A Forum <https://forums.att.com/t5/custom/page/page-id/search-page>

Other Guides by Shelby Johnson & Tech Media Source

Amazon Fire TV User Manual: Guide to Unleash Your Streaming Media Device

Kindle Fire HDX & HD User's Guide Book: Unleash the Power of Your Tablet!

Kindle Paperwhite User's Manual: Guide to Enjoying your E-reader!

Apple TV User's Guide: Streaming Media Manual with Tips & Tricks

iPad Mini User's Guide: Simple Tips and Tricks to Unleash the Power of your Tablet!

iPhone 5 (5C & 5S) User's Manual: Tips and Tricks to Unleash the Power of Your Smartphone! (includes iOS 7)

Facebook for Beginners: Navigating the Social Network

How to Get Rid of Cable TV & Save Money: Watch Digital TV & Live Stream Online Media

Chromecast User Manual: Guide to Stream to Your TV (w/Extra Tips & Tricks!)

Google Nexus 7 User's Manual: Tablet Guide Book with Tips & Tricks!

Roku User Manual Guide: Private Channels List, Tips & Tricks

www.techmediasource.com

www.ingramcontent.com/pod-product-compliance
Lightning Source LLC
Chambersburg PA
CBHW060630210326
41520CB00010B/1543